THE SCIENCE OF SCIENCE

A guide to thinking like a scientist

Crystal C. Li

Table of Contents

Introduction   1

Part I: What is science?   2

Part II: The scientific method   7

Part III: Scientific vocabulary   36

Part IV: Sharing discoveries   52

Part V: Science is great, but not perfect   60

INTRODUCTION

Welcome to *The Science of Science*! During your lifetime, you will learn countless scientific facts, inside and outside the classroom. Those facts are important, but it's even more valuable for you to understand how those facts (and others like it) were discovered and to be able to think scientifically on your own. That's what this book is here to help with.

This book won't give you a bunch of science facts to memorize, but it will teach you about science itself: what it is, why it's useful, how it works, and important ideas to keep in mind when learning science.

PART I
WHAT IS SCIENCE?

What do you think of when you hear the word "science"? Maybe you think of someone in a laboratory mixing chemicals together. Maybe you think of experiments where things explode. Maybe you think of tiny atoms and molecules.

These are ideas that most people relate to science, but science isn't all about chemicals and explosions. There are so many things that count as science because science is everywhere around us; it's the study of the natural and physical world. Scientists learn and make discoveries by asking questions and then using observations and experiments to answer them (and to find more questions to ask) in a process called **research**.

What makes science science is that it is based on what we can *observe* (see, hear, or otherwise perceive). Scientists draw logical conclusions using that observational evidence. They don't rely on gut feelings or what "feels right."

Of course, there are things that scientists cannot see, such as gravity or magnetic fields, but can logically conclude that they exist because of what we *can* observe: we see objects fall to the ground and we see magnetic objects attract and repel.

Anything that can be observed can be studied using science. Maybe you want to know whether you can run faster on an empty stomach or after eating a snack. If you test it out to find the answer, that's science!

There are so many topics that scientists are studying: how to cure diseases, how animals communicate, how to generate and transmit electricity, how volcanoes erupt, how to make plants grow faster, how the universe is expanding, and much, much more!

Here are just a few of the many branches of science:

Biology: the study of life—specifically, the origins, properties, and processes of plants and animals. Thanks to an understanding of biology, we can cure diseases and plant food crops.

Chemistry: the study of the substances that the world is made out of and how these substances combine, react, and change. Chemistry has helped us create stronger materials such as steel and allows us to clean our water and air.

Physics: the study of mechanics (how and why objects move and behave across space and time), energy, electricity, and magnetism.

Electricity and magnetism have helped develop motors to drive our cars and generators to light up our buildings. And without understanding mechanics, we wouldn't even have cars or sturdy buildings in the first place!

Geology: the study of Earth's physical structure and makeup (such as its rock layers) and how they change over time. Topics of study here include mining, earthquakes, volcanoes, the formation of mountains, and the rock cycle.

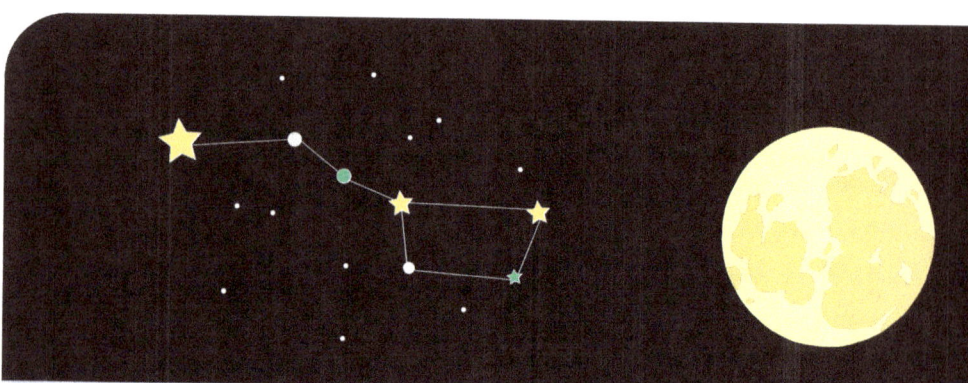

Astronomy: the study of the universe, outer space, and the celestial objects in it (such as stars and planets). Astronomers try to answer questions such as "How fast is the universe expanding?" and "How long would it take light to travel from a distant star to Earth?"

Another way to think of science is that it's a method for updating and adding to what we know about the world. Importantly, what we know about the world continuously changes as scientists make new discoveries.

This book is all about how scientific discoveries are made and what you should know as a consumer (and maybe future discoverer!) of science knowledge. It will help you become a better scientific thinker.

PART II

THE SCIENTIFIC METHOD

Scientists conduct research by using the components of the **scientific method**, a logical and systematic process for finding answers. Before we walk through it, keep in mind that not all scientists follow these steps in this exact order because everyone has a method that works best for them. However, each part of this process is important.

> **Takeaway #1: Science research uses the scientific method, a process of observation and experimentation, to answer research questions.**

First, we notice something we want to know more about. Maybe you notice that on some days you are able to run faster than on other days. This step is called **observation**.

Observation means carefully examining something and noticing its features and properties. We can observe with our eyes and ears, or with tools such as telescopes and microscopes.

After we notice something, we usually want to study it to learn more. That's why the next step is to ask a **research question** about it.

If you can run fast on some days but not others, you will probably want to know why. Our research question can then be "What causes people to run faster?" Now we have a goal for our research: to answer our question.

What we find out in our research should answer our question, and we can now plan our experiment or study in a way that will do just that.

Then, we form a **hypothesis**. That is, try to answer your own research question. A hypothesis is a prediction about the answer to your research question that you can test out.

Maybe you think that the reason why you can run faster some days is because you ate a snack before you started running. Your hypothesis would be "If people eat a snack, then they can run faster afterwards."

Now we need to test our hypothesis.

There are many ways to do this, but one option is to conduct an **experiment**. In an experiment, we try to see how one thing (an independent variable) causes a change in something else (a dependent variable). In this case, we are investigating how eating a snack causes a change in how fast people can run.

Experiments

Let's say we're conducting an experiment this week. Here are the **procedures** of it: First, you gather a group of classmates to study and divide them into two groups. (Don't tell them your research question or hypothesis because it might change how they act in the experiment.)

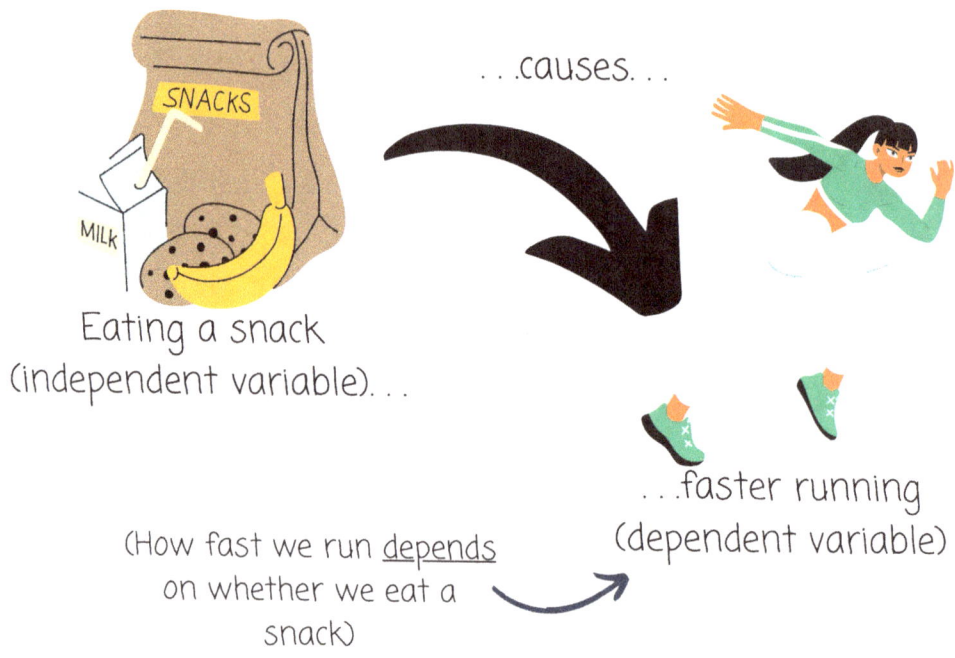

Eating a snack (independent variable)... ...causes... ...faster running (dependent variable)

(How fast we run <u>depends</u> on whether we eat a snack)

The first group doesn't eat before running. You just go ahead and time how long it takes each person to run across the field at school on a timer. (The less time, the faster they are.) The amount of time it takes everyone to run will be the **data** you collect (the information you will later use to draw a conclusion).

You have the second group of classmates eat a granola bar fifteen minutes before running. Then you use a timer again to time how long it takes each person in this group to run across the field.

Then you find the average amount of time it took people in Group 1 to run and the average time it took people in Group 2, and compare the two numbers. (*Average means add up the times for everyone in one group and divide by the number of people in the group*)

If Group 2's time is shorter than Group 1's, that's **evidence** that backs up your hypothesis. A snack really might make people run faster. If the two groups have about the same average time or if Group 1's time is shorter, then you don't have evidence for your hypothesis.

In this experiment, the factor that we think causes a change (whether or not people eat a snack) is the **independent variable**, and what we think is being changed as a result (how fast people run) is the **dependent variable**: the outcome we are measuring.

What are some different types of measurements we can take? Size, time, weight, temperature, and even customer satisfaction with star ratings!

The independent variable is the one thing that we as scientists directly change (**"manipulate"**) in experiments. It's what makes experiments so useful: since we are changing one variable, we can show whether that variable causes a change in another variable. Therefore, we can show whether there is a *cause and effect relationship* between the two.

The reason why we have two groups in the experiment (Group 1 and Group 2) is because we want to compare the speeds of the two groups to see if eating a snack made a difference.

This way we have two different conditions of the independent variable:

one where people eat a snack, and one where they don't.

The one where your classmates don't eat a snack is called the **control group**—that's what Group 1 is.

OK. Let's say Group 2 really turned out to have a faster average speed (shorter time) than Group 1. What does our experiment tell us? Well, we have evidence to support our hypothesis, but that doesn't mean we're *definitely* correct. Maybe it was a fluke! Maybe we got lucky that Group 2 just happened to run faster.

To be more sure of our hypothesis, we need more evidence. The way to get more evidence is to repeat our experiment. This is called **replication** because we are trying to get the same results again and again.

The more times we repeat our experiment and get the same results, the less likely it is that we got the results by random chance.

Takeaway #2: Scientists need to replicate (repeat) studies many times in order to check whether they continue to get the same results.

If Group 2 ran faster than Group 1 this week, that might just be luck. But if the same thing happens again five weeks in a row, then it's probably not luck; very likely it's because Group 2 ate a snack but Group 1 didn't, just like we expected.

After repeating the experiment many times, on many different groups of classmates, let's say you find that in *most* cases, the group that eats a snack runs faster than the group that doesn't. We conclude that eating a snack helps run faster.

The thing is, on a few replications of the experiment, the group that doesn't eat runs faster and sometimes the two groups are pretty much tied.

This tells us that food isn't the only **variable** causing people to run faster or slow: there must be factors involved.

Maybe some people are taking a different path across the field: some are running straight and others are running a bit diagonal. This would mean some are running longer distances than others.

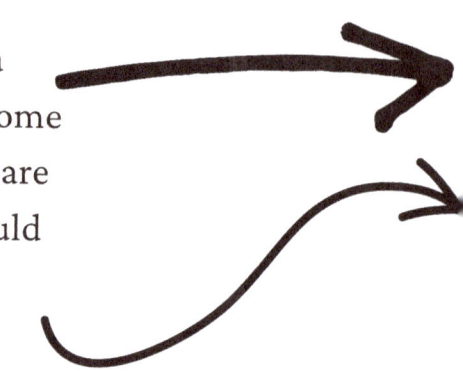

Maybe one group is running in the morning and the other group is running in the afternoon.

Maybe some people wore more comfortable shoes that are suited to running while others didn't.

All of these variables might affect the time it takes people to run across the field, and it might be hard to tell whether your classmates in Group 2 are running fast because of the snack or because of these other causes.

We need to keep all these other variables the same (or **constant**) in and across Groups 1 and 2—make them **controlled variables:**

 You should give your classmates an exact starting point to run from and stopping point to run to so that everyone runs the exact same distance.

You should have the two groups run at the same time of day (either morning or afternoon).

You should ask your classmates to wear comfortable clothes and sneakers when running.

After controlling these variables (and others that you can think of), if Group 2 again runs faster than Group 1, you can be more confident that it's because of the snack and not because of other factors such as distance, time of day, and shoes.

Scientific sidenote: Don't confuse *controlled variables* with a *control group*! In a well-designed experiment, we need both.

A **control group** (or control condition) is the group of things or people being studied in an experiment that is *not* given a treatment—it's in the natural, untouched state. (Think of Group 1 in our study, which didn't receive our "treatment": a snack.)

If a scientist is trying to see if a sleeping pill works, he or she will use at least two groups of people: one group will be given the sleeping pill and the other group will be given no pill (or a fake pill that has no effect). The group that is not given the sleeping pill is in the *control condition*, while the one that is given the pill is in the *treatment condition*. The scientist will see whether the people in the treatment condition fall asleep quicker than the people in the control condition (who got no sleeping pill).

Thus, a control condition/group is useful because scientists can *compare* the results of the treatment condition with it to see whether the treatment was effective.

Controlled variables are factors that we keep the same *across* our treatment and control conditions. (Think of distance, shoes, and time of day in our study.)

We're not trying to study these variables, but they may influence the variables we *are* studying (the independent and dependent variables) if we don't keep them constant, and we don't want that.

If we *do* keep them constant, we can conclude that a difference we observe in our dependent variable between the treatment and control conditions was most likely caused by the independent variable because we made sure that the independent variable was the only thing that was different between the control condition and treatment condition. The controlled variables couldn't have caused the change because they were the same in both conditions.

Now that you've done what you can to control variables such as the time of day the groups run and what people wear while running, let's say that the next few times you repeat the experiment, you find that almost every time, the group that eats a snack runs faster than the group that doesn't.

That makes for pretty good evidence that the granola bar snack is helping people run faster!

Now, maybe you want to find more answers to your research question ("What causes people to run fast?"). You might think, what if, instead of a granola bar, you had your classmates eat something else? Would that change how fast they can run?

Some snacks might be better than others at providing an energy boost. If you can find out which kinds of foods are more helpful than others, then you will have a more precise answer to the research question.

We should go back to the hypothesis step and think of other foods that might make us run faster or slower than a granola bar does.

Maybe the granola bar has more sugar and makes you feel more full than an apple does. Because of this, you might think a granola bar might give people more energy than an apple. The new, refined hypothesis can then be "If my classmates eat an apple before running, they will run slower than if they ate a granola bar, but still faster than if they ate nothing."

As you can see, we keep forming hypotheses and testing them until we have answered our research question as best as we can. If our first hypothesis turns out to be wrong based on the data we collect, then we would change our hypothesis or make a new one and then continue gathering evdience.

In this way, the scientific method is one big *cycle*. We repeat the steps again and again to get more precise answers.

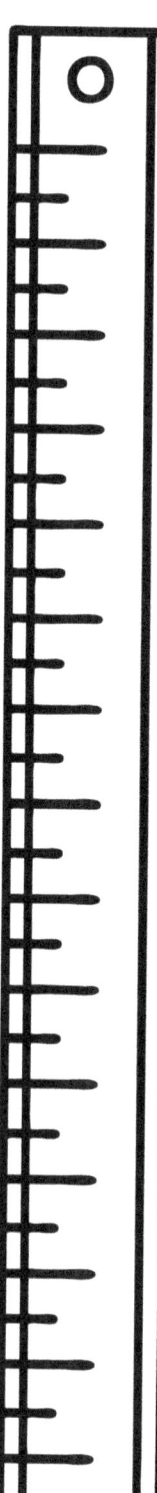

Correlational/Observational Studies

Now, let's say we want to test a new hypothesis: "If a person is taller, then they will run faster."

The problem here is that you can't tell some people to be taller and other people to be shorter, which is what you have to do if you want to test this hypothesis experimentally (remember that in an experiment, we as researchers have to change the independent variable). You can't control the height of your classmates in the same way you can control whether or not they eat a snack. That's why we can't use an experiment to test this hypothesis.

What can we do? Another option is to gather a bunch of classmates who are different heights and see if the taller people can run faster than the shorter people.

This is called a **correlational** or **observational study** because we are trying to see if there is a relationship (correlation) between these two variables, which we don't manipulate but simply observe. We expect that as height increases, time to run decreases.

> The data we need from each person are their height and the amount of time it takes them to run across the field.

Scientific sidenote: Correlations

If values of one variable increase as values of another variable also increase, that is called a **positive correlation**. If values of one variable decrease as values of another variable increase, that is called a **negative correlation** (the variables change in opposite directions). Can you think of more examples of positive and negative correlations?

Here are a few ideas.

Positive correlations:
- The *taller* a mountain is, the *more* snow there is at the top of it.
- Basketball players that spend *more* time practicing and training score *more* baskets during games.

Negative correlations:
- As a person's age *increases*, the speed at which they can learn new languages *decreases*.
- Temperatures *increase* as the distance between a location and the equator *decreases*.

Here is something important to always remember: *Correlation does not mean causation!* Just because there is a relationship or correlation between two variables doesn't mean that one *causes* the other.

There could be additional variables that are influencing both of the variables we're looking at.

For example, if ice cream sales are increasing at the same time that the number of shark attacks increases, does that mean ice cream sales are *causing* shark attacks? Probably not! The reason they are related is likely because a *third variable* (a **confounding variable**) is influencing both variables: <u>temperature</u> (time of year). As temperatures increase when we move closer to summer, more people want ice cream *and* more people are at the beach, increasing the chances of a shark attack.

Another reason why correlation doesn't mean causation is because the relationship between the two variables might be a complete coincidence!

However, some variables that are correlated might really have a cause-effect relationship, like the basketball example. It makes complete sense that more practice causes a player to score more baskets in a game—but just remember that correlational data alone showing that the players who practice longer are the ones who score more baskets is not strong enough evidence to support that claim. We'd need an experiment where we make different players practice for different amounts of time in order to show a cause and effect relationship.

Takeaway #3: Correlation does not mean causation. Experiments are the only type of study that can show a cause-and-effect relationship between two variables.

Okay, going back to our study on running: we will be able to see whether the people who run fast tend to be taller than the people who run slower, but we won't be able to see whether being taller *causes* people to run faster. But this is the best option we have since we can't change people's heights at will, so we'll use it!

Remember: It's not always possible to run an experiment because it's not always possible to change certain variables.

Let's say we collect this data:

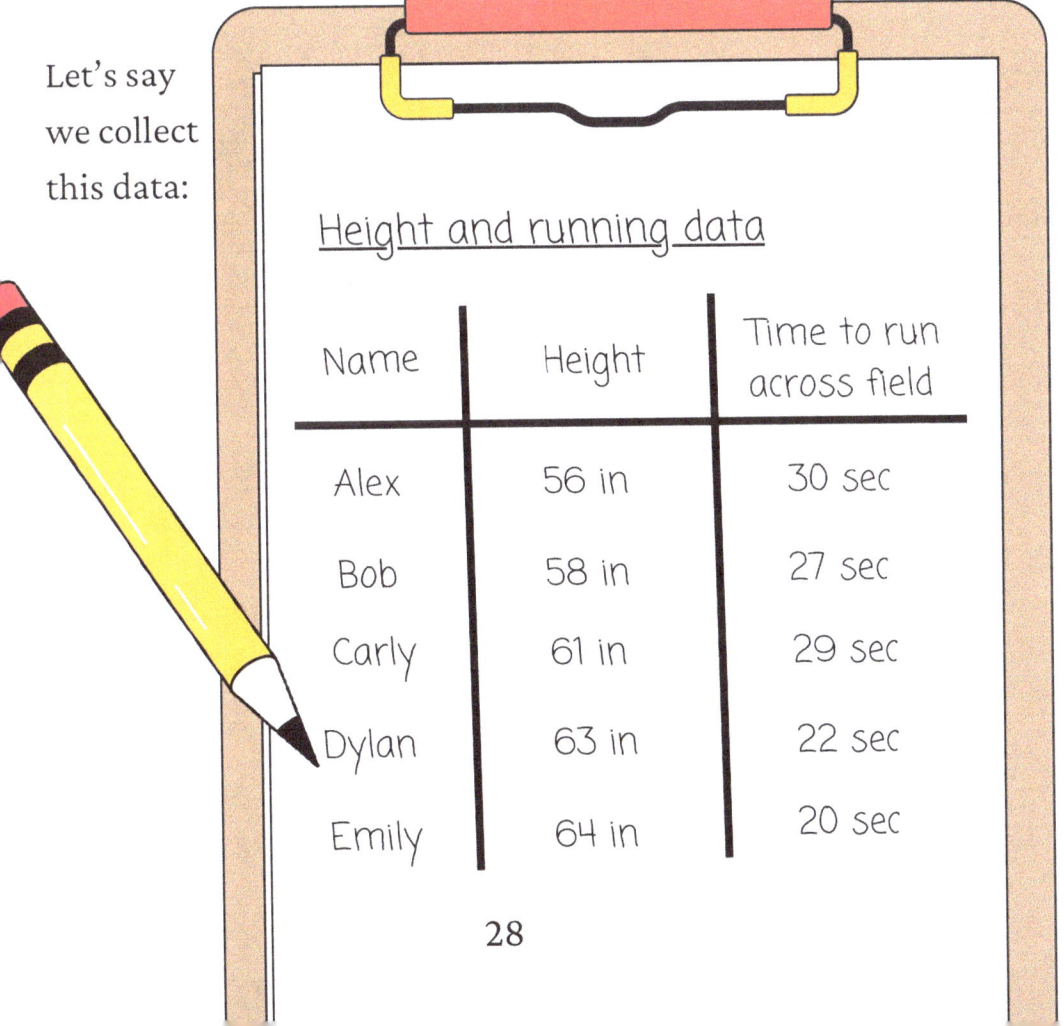

Height and running data

Name	Height	Time to run across field
Alex	56 in	30 sec
Bob	58 in	27 sec
Carly	61 in	29 sec
Dylan	63 in	22 sec
Emily	64 in	20 sec

What do you notice? For the most part, the taller people, like Dylan and Emily, seem to take less time to run across the field than the shorter people, Alex and Bob.

There are exceptions (Carly is taller than Bob but it took her two seconds longer than Bob to run across the field), but in general, the pattern is that the taller someone is, the faster they run (remember, shorter time means faster).

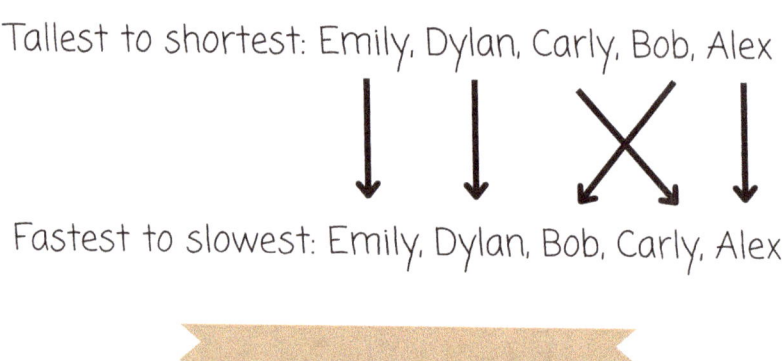

Scientific sidenote: Scatterplots

If we plot each person as a point on a graph, they will roughly form a downward-sloping line, which tells us that it's a negative correlation. (If it were a positive correlation, the points would form an upward-sloping line like in the second example graph on the next page.)

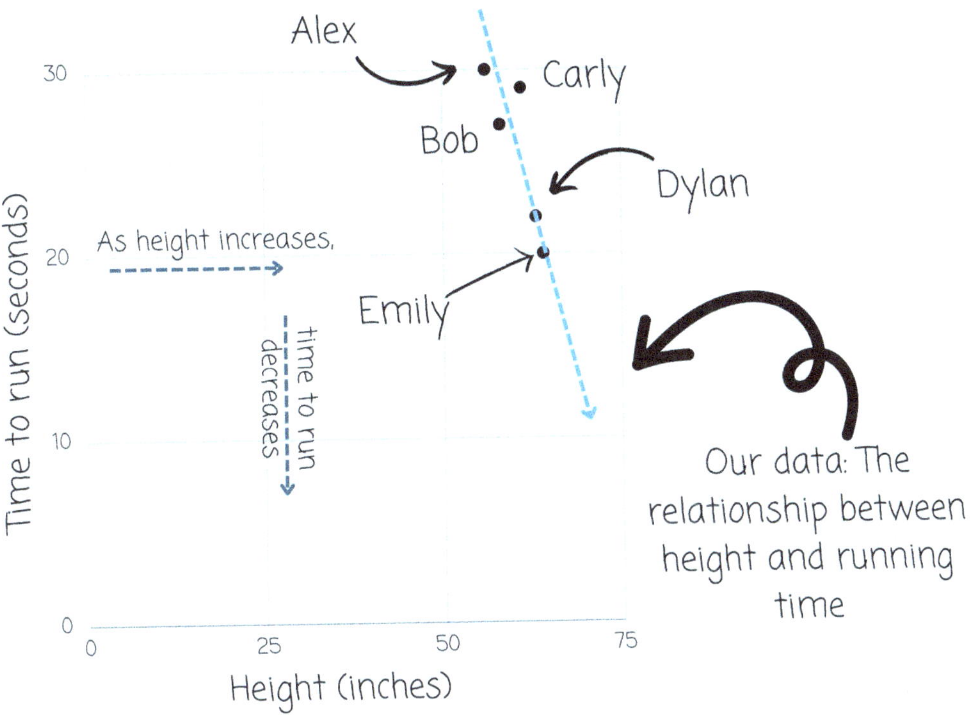

Our data: The relationship between height and running time

Scatterplot with axes "Height (inches)" (x-axis) and "Time to run (seconds)" (y-axis), showing points labeled Alex, Bob, Carly, Dylan, Emily. Annotation: "As height increases, time to run decreases."

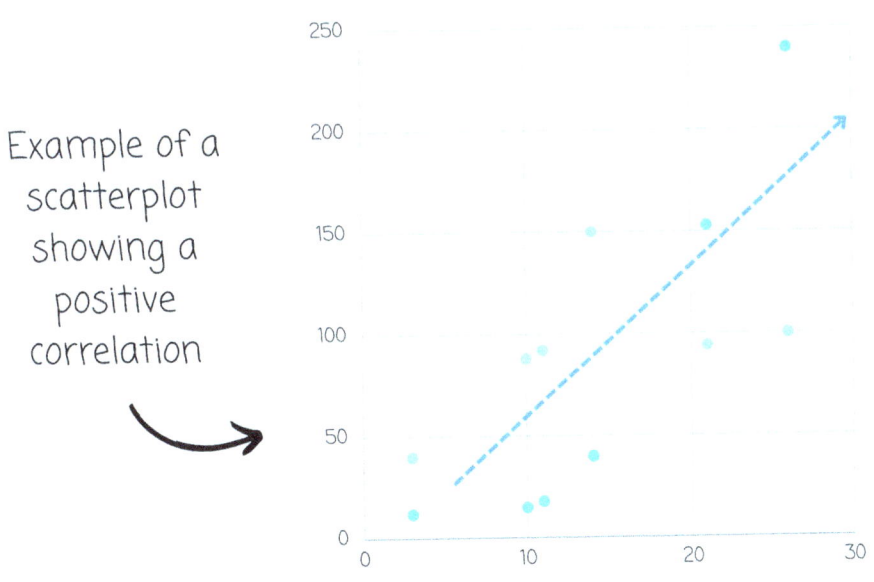

Example of a scatterplot showing a positive correlation

This data support our hypothesis partially, but not fully, because like we talked about, we aren't able to show that increasing someone's height directly causes them to run faster.

Our correlational data is pretty good evidence though! We can use our own reasoning to make an argument about whether there might be a cause-effect relationship. Here, we can argue that even though we don't have experimental evidence, it makes a lot of sense that greater height causes people to run faster because they have longer legs to propel themselves forward with more power.

At the same time, however, there could be **c**onfounding variables (factors that affect *both* height and running speed). Take nutrition, for example: the more nutrition someone has, the taller they will probably grow; at the same time, the more nutrients their body has, the more strength they have to run fast.

We don't know whether the reason why height and speed are correlated is because a third variable relates to both of them or because height really does increase speed.

And even if there is a cause-effect relationship, we also technically can't be sure whether it's taller height that causes faster running, or faster running that causes people to grow taller. Who knows? Maybe exercise helps you grow taller!

If you wanted to learn even more about what makes people run faster, you could make even more hypotheses and keep testing them out. Maybe you think that training and practice can allow shorter people to run just as fast as (or even faster than) tall people, and you can test this out by gathering a group of shorter classmates to practice running for a few weeks and see how they do compared to tall classmates who didn't practice.

Again, the scientific method is an endless cycle of asking questions, making predictions, testing the predictions to answer the questions, and then either making more predictions, adjusting our first prediction (if our data and observations don't support it), or asking more questions.

Scientists use this method with anything they want to study, from tiny insects to distant galaxies.

Scientific sidenote: Experiments in different fields of science

Experiments in biology:
Biologists might want to know whether a medication called an antibiotic is effective in fighting bacteria, so they collect bacteria, expose half of it to the antibiotic and don't do anything to the other half, and then see if the first half ends up getting killed.

Experiments in physics:
To see if gravity pulls all objects down to Earth with the same force, some physicists sucked all the air out of a massive container and then dropped a feather and a bowling ball in it at the same time, from the same height. Without any air particles to resist the downward motion of the feather and slow it down, the bowling ball and the feather hit the ground at the exact same time. This showed that gravity acts with the same amount of force on all objects on Earth, no matter their weights and sizes.

Experiments in astronomy:
How can scientists possibly experiment on faraway stars and galaxies? Oftentimes astronomers rely on observation

rather than experiments to study outer space. But experiments have still been used in astronomy!

If astronomers want to know how far away a star is, they can find out by measuring how bright the star looks from Earth using special sensors and using that brightness to calculate the distance. They can do this because there's a mathematical relationship between how bright a star looks and how far away it is. And the reason why we know of this relationship is because of <u>experiments</u>: scientists varied the distance between a brightness sensor and a light source and saw how increasing that distance caused the sensor to sense the light source as less bright.

Can you think of other experiments that scientists (or you!) could do to answer a research question? Try to think of different cause-and-effect relationships that you could test out...

PART III

SCIENTIFIC VOCABULARY

Many commonly used words in science are misunderstood. Two examples of this are "theory" and "fact." Let's make sure you know how to use these words correctly.

Theory

In everyday non-scientific conversation, you might use the word "theory" to mean that you have a "hunch" or "guess" about something. Maybe you have a substitute teacher one day and you whisper to your friend that you have a "theory" as to why your regular teacher isn't there that day. You don't have evidence, but you have a reasonable guess. This use of the word "theory" in normal conversation causes many people to mistakenly believe that a scientific theory is only a guess.

In reality, a scientific theory is an explanation about observed **phenomena** (a scientific word for "things," "processes," or "events") in the universe that logically ties many related facts and pieces of evidence together. Theories are based on evidence gathered using the scientific method and are widely agreed on.

For example, the **theory of evolution** explains how species of plants and animals change over time: their genes change generation by generation to adapt to their environment and increase chances of survival.

This theory was based on scientist Charles Darwin's observations of a type of bird called **finches**.

He noticed that the insect-eating finches had longer, pointier beaks to catch insects easily by wood-pecking, and that these finches were less likely to starve than the finches without these helpful wood-pecking beaks.

Darwin theorized that in the past, some finches had longer, narrower beaks and others had shorter ones, but the finches with the longer beaks survived while the finches with shorter ones starved to death.

Because of this, the long-beaked finches were the ones that lived to have offspring, *passing down the genes for their beaks to the next generation*. Therefore, the next generation of finches had mostly long beaks instead of short beaks. Over the course of many many years, nearly all newborn finches in the species began to have these long beaks suited to wood-pecking and eating insects.

According to Darwin, going from short beaks to long beaks was an example of evolution for these finches.

Check out the next page for an illustrated explanation.

Darwin's Theory of Evolution

Year 1: Due to random genetic variation, some birds have long beaks and others don't.

Year 3: Many of the short-beaked finches starve, while the long-beaked ones live to lay eggs and pass their genes down.

Year 5: the next generation has more long-beaked finches than short-beaked ones.

Year 100: All birds in this species are now born with long, pointy beaks because generation by generation, the number of long-beaked finches increases while the number of short-beak ones decreases.

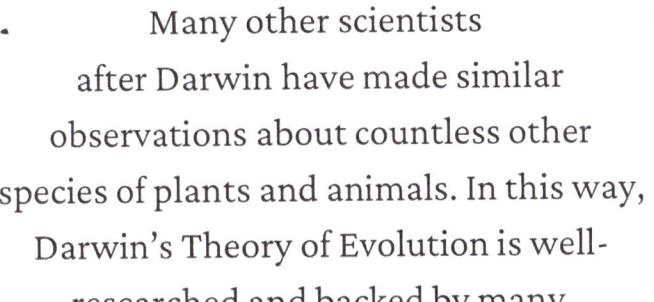

Many other scientists after Darwin have made similar observations about countless other species of plants and animals. In this way, Darwin's Theory of Evolution is well-researched and backed by many years of evidence.

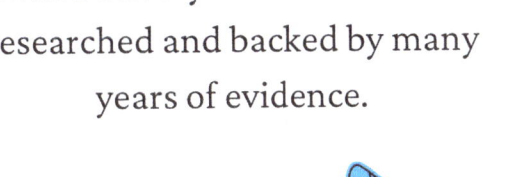 

Takeaway #4: A scientific theory is not just a guess. It's an explanation or description of something about the world that is based on a large amount of evidence.

Fact

What about the definition of a scientific fact? That seems pretty straightforward: a fact is something that is true. But we need to consider what makes something true. How much evidence do we need to have in order for something to be considered a scientific fact?

An observation or piece of information in science is a fact when all of the available evidence and all the information we have shows that it is true. Usually (but not always), this means that when something is a fact, everyone or almost everyone agrees on it.

Proof

We feel certain about facts because they're backed by evidence, not because they're *proven*. What's more, proof is a word that should never be used in science. You may think that proof is just another word for evidence, but the two actually have different meanings.

Evidence is information that can be used to support a scientific claim. The logic is: "a possible reason why this evidence/information exists is because my claim is true." But just because you have evidence doesn't necessarily mean the claim is correct.

Of course, the more evidence you gather that supports (and doesn't contradict) your claim, the more likely your claim is to be correct. You'll just never know for sure because one day you might find evidence that *does* contradict your claim.

On the other hand, if you have proven a claim, that means you have information showing that the claim is definitely 100% true and there is absolutely no way for it to be incorrect.

Unlike with evidence, the logic with proof is: "if this information exists, then my claim *must* be true."

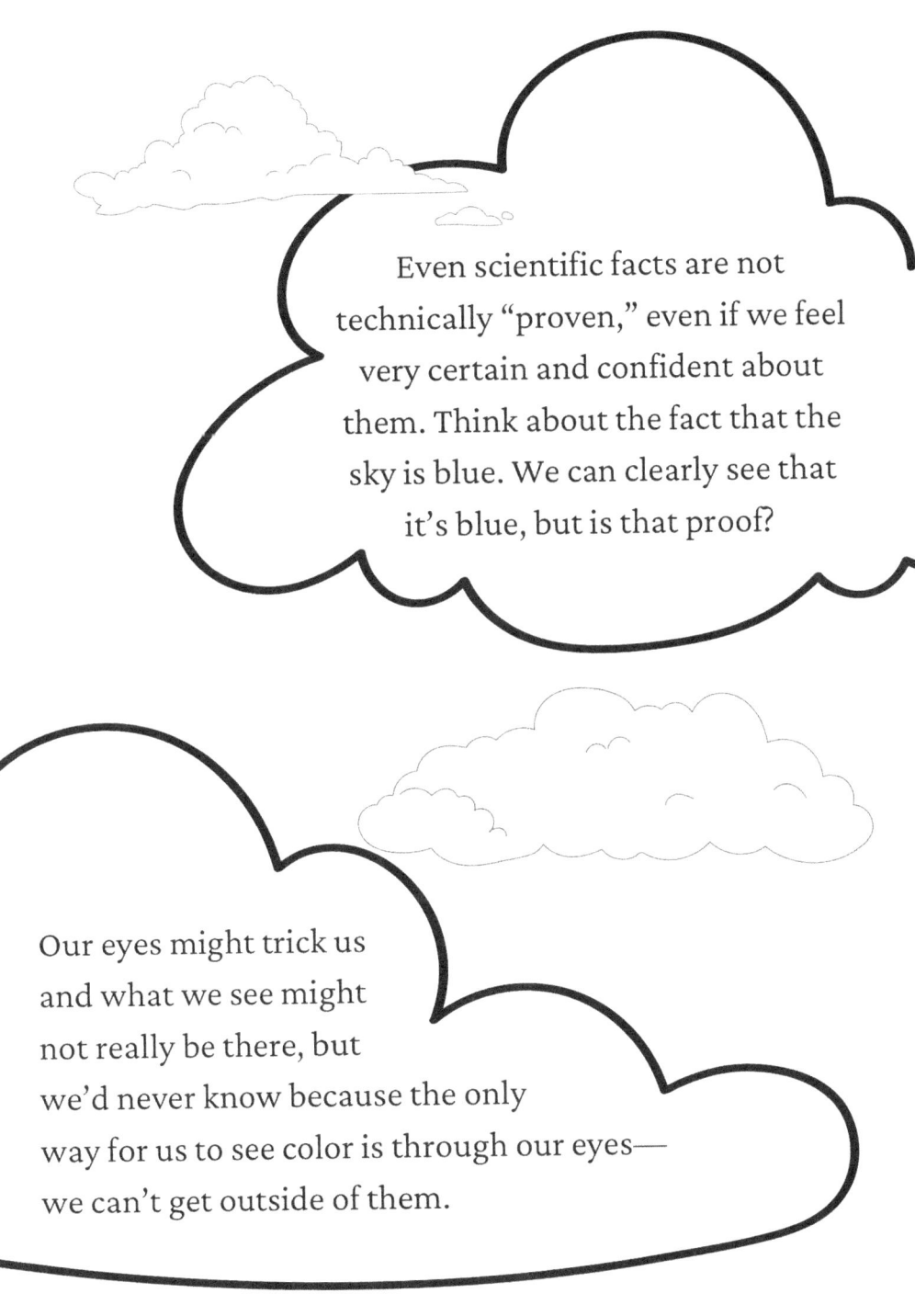

Even scientific facts are not technically "proven," even if we feel very certain and confident about them. Think about the fact that the sky is blue. We can clearly see that it's blue, but is that proof?

Our eyes might trick us and what we see might not really be there, but we'd never know because the only way for us to see color is through our eyes—we can't get outside of them.

Takeaway #5: There is no such thing as proof in science, only strong evidence.

The point isn't that you should start doubting everything just because proof doesn't exist; that's unproductive. Knowing that proof doesn't exist in science should simply remind you to be open-minded about scientific ideas changing over time or one day being falsified (shown to be incorrect) as more evidence is gathered. If nothing is proven, then nothing is completely set in stone.

Now let's talk more about evidence because it's the backbone of scientific knowledge.

A few words about evidence

Something important to remember is that when you have a scientific claim or hypothesis that you want to research, you should always welcome evidence that doesn't support (or contradicts) what you believe.

If you only look for evidence that supports your hypothesis, then you might miss important information that leads to a completely different conclusion. The last thing you want is to be completely wrong and not even know it! If scientists did this, then they wouldn't really be looking for the truth—they'd just be trying to show that they're right. This is called **confirmation bias**: only looking for evidence that supports our existing beliefs and ignoring evidence that doesn't.

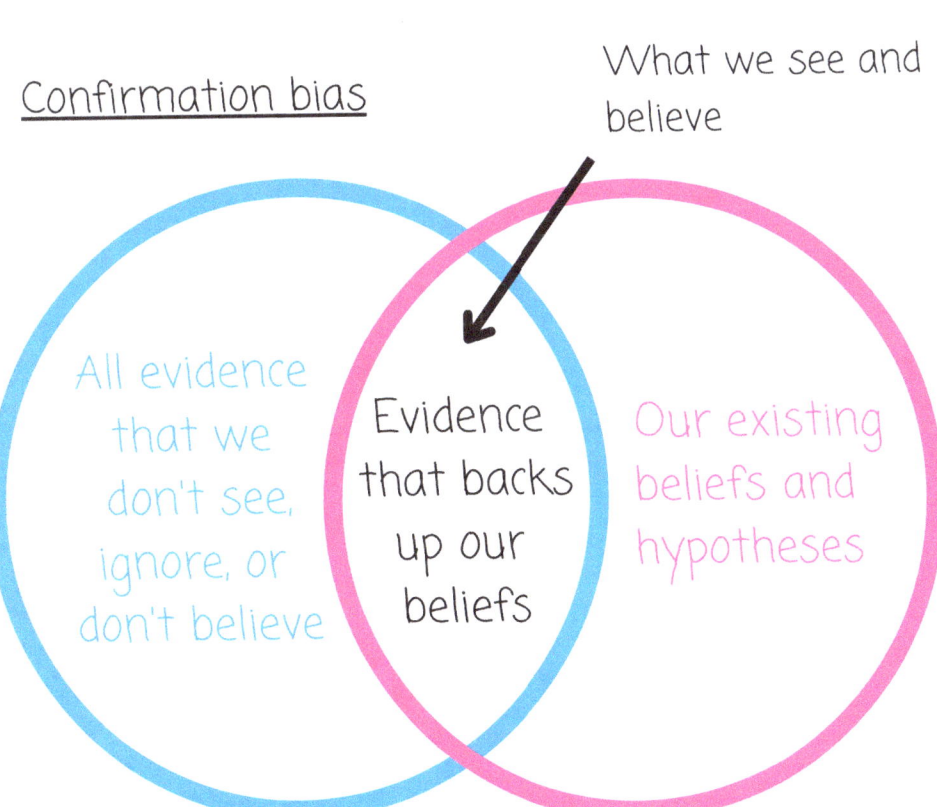

Takeaway #6: Beware of confirmation bias, the temptation to only pay attention to evidence that supports your expectations while ignoring evidence that says you're wrong.

Everyone is guilty of it at times, but knowing that it happens can remind us to veer away from it.

We can do this by, for example, doing thorough background reading on a wide range of topics related to our research question, making sure we become aware of theories and past findings that say we may be incorrect.

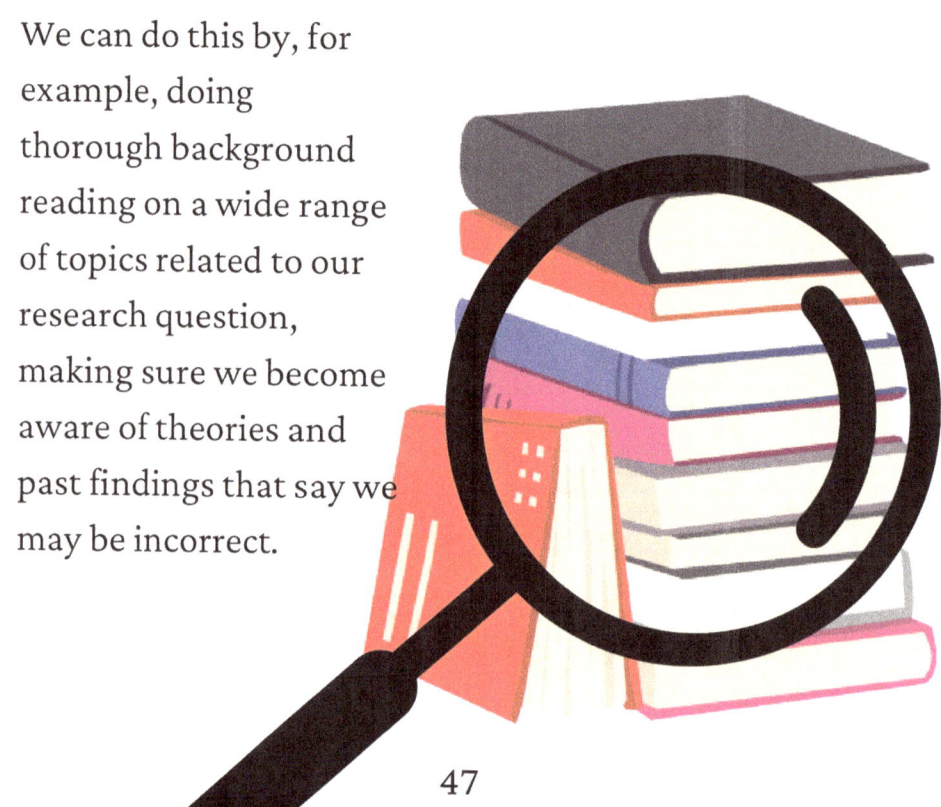

The ability for confirmation bias to lead us away from the truth is something you should keep in mind whenever you're looking for answers.

> When you're searching online, check many sources. Don't just stop at the first article, website, or post that answers your question or that confirms what you think the answer is. Be open to information that says the opposite of what you believe and of what the first answer you found says.

Likewise, when you're doing science research, don't just stop once you find evidence that supports your hypothesis. Keep repeating trials and collecting as much data as reasonably possible before drawing a conclusion, even if the evidence you find isn't what you want or expect to see.

Another way to combat confirmation bias in science is **peer review**. You might be told to proofread your classmates' writing so that everyone can help one another check for mistakes. Scientists do the same thing!

Before a paper about a scientist's discoveries is published, other scientists usually read it to make sure the study was carried out properly and in an **unbiased** way. Research is biased if, for example, researchers only collect data that support their hypotheses.

Scientists also double-check one another's discoveries by replicating one another's studies and experiments. If replications yield the same results as the original, then that's good reason to trust the findings. Other times, some scientists might find evidence that contradicts another scientist's findings. If that happens, then more research is needed to clarify what is actually true.

Scientific sidenote: Debates in science

Because different researchers may find different evidence, have different interpretations of evidence, and therefore draw different conclusions, scientists can sometimes disagree on new discoveries. These disagreements can usually be sorted out once more evidence is collected and more analysis is done, but this takes a lot of time (sometimes many years), so we have to be patient. Science can't always give us immediate answers!

On most issues and discoveries, however, scientists end up being able to resolve their disagreements by the time they present their research to the public. (If not, then it's *your* job as a consumer of science knowledge to make a judgment about what is most likely true. Use what you have learned and will continue to learn in this book about good scientific practices and study designs to evaluate the research and claims of scientists!)

> **Takeaway #7: Science is a collaborative process. Scientists work together on research projects and double-check one another's work to make sure they produce reliable findings.**

At this point, we've covered a lot about how science research is done. The next step is to share the research findings with other scientists and with the public.

PART IV
SHARING DISCOVERIES

After scientists make discoveries, they want to communicate them to the public because their findings could be beneficial for people to know. This is just as important as making the discoveries themselves because discoveries would be useless if no one ever learned about them!

> There are a few things you should know about how results of scientific research are reported so that you can both understand and communicate scientific knowledge yourself.

A lot of times visual displays of data are the most effective way to summarize and communicate to an audience what the results of a research study are.

Let's first look at **bar charts**. These are used to *compare* multiple groups (here, we're comparing the the height of trees in three different locations).

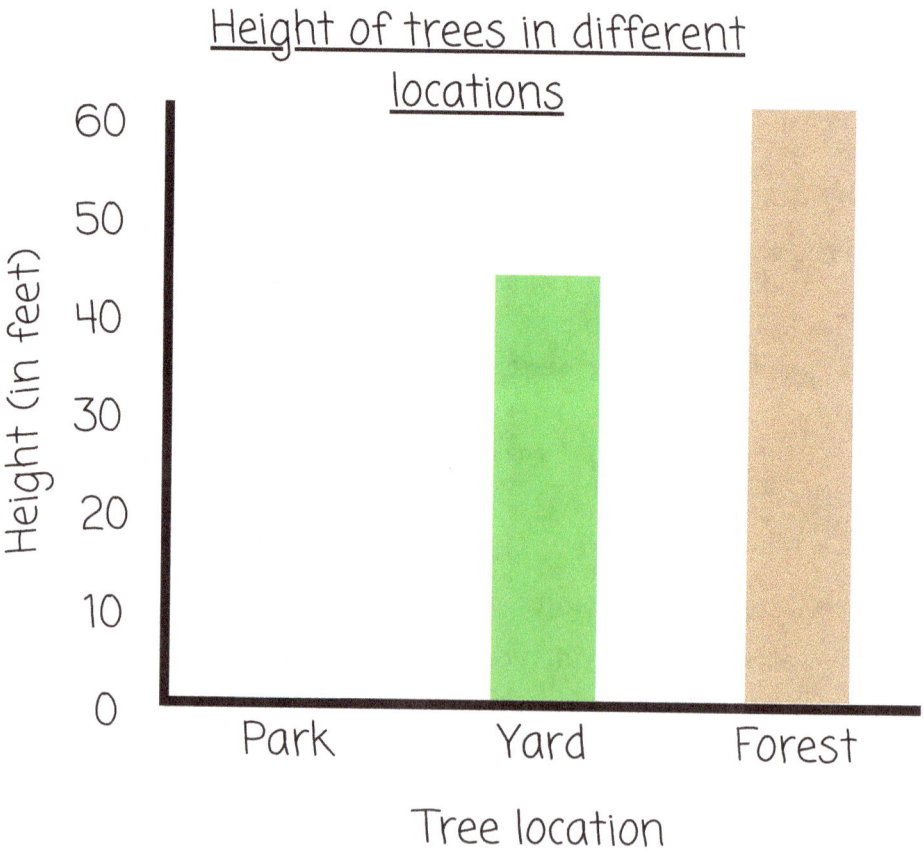

The bars show that the tree in the forest has grown the tallest, followed by the tree in the park, and then the tree in the yard.

Here's another bar chart with the same information but presented differently. Do you notice anything about it that might be misleading and trick people into believing something that isn't accurate?

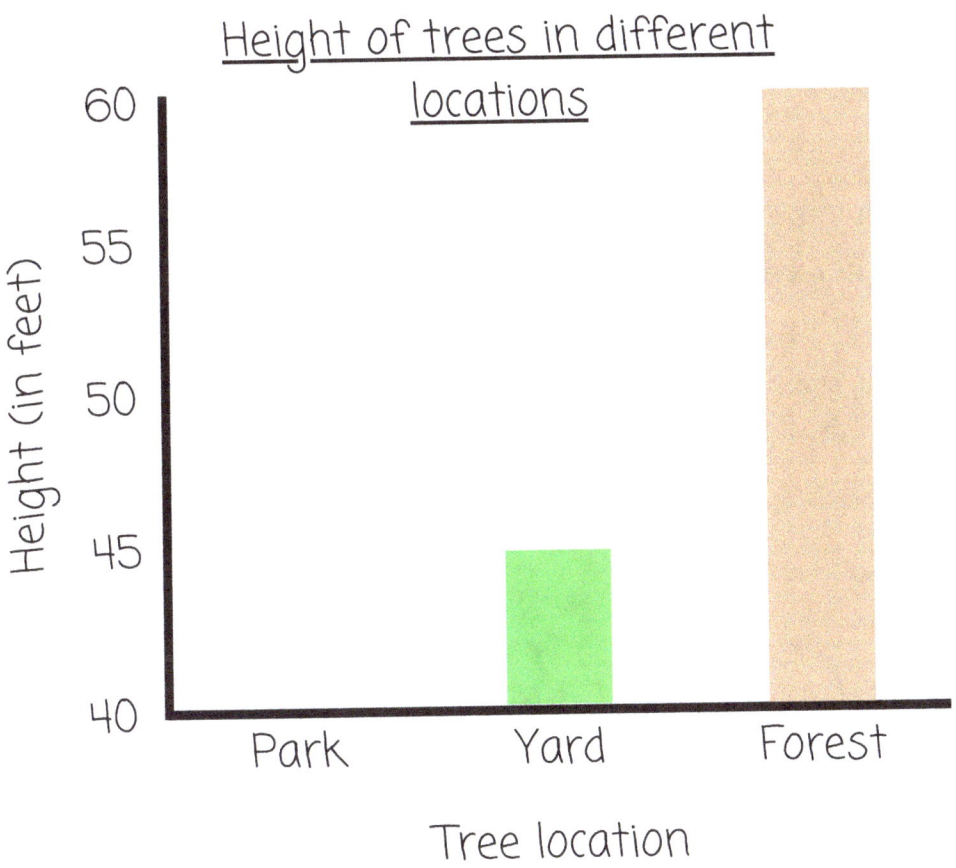

Notice that the **scale** on the vertical side of the chart (the **y-axis**) doesn't start at zero. It starts all the way at 40! This makes the heights of the bars look way different from the first bar chart.

Based on the height of the bars in the second one, it seems like the forest tree is twice as tall as the park tree, and the park tree is twice as tall as the yard tree.

In reality, the forest tree is only 10 feet taller than the park tree and the park tree is only 5 feet taller than the yard tree!

Remember to be careful when reading charts like this. Notice everything: the labels under the bars, the heights of the bars, and the numbering on the y-axis.

Bar charts are supposed to illustrate and compare the true values or measurements of multiple categories. **Therefore, it's always best to label the y-axis starting with zero.** That way, the bars give an accurate picture of how big, expensive, frequent, fast (or whatever) something is.

55

Line graphs are also common data displays, and they are useful when scientists want to show a *trend, pattern,* or *change over time.*

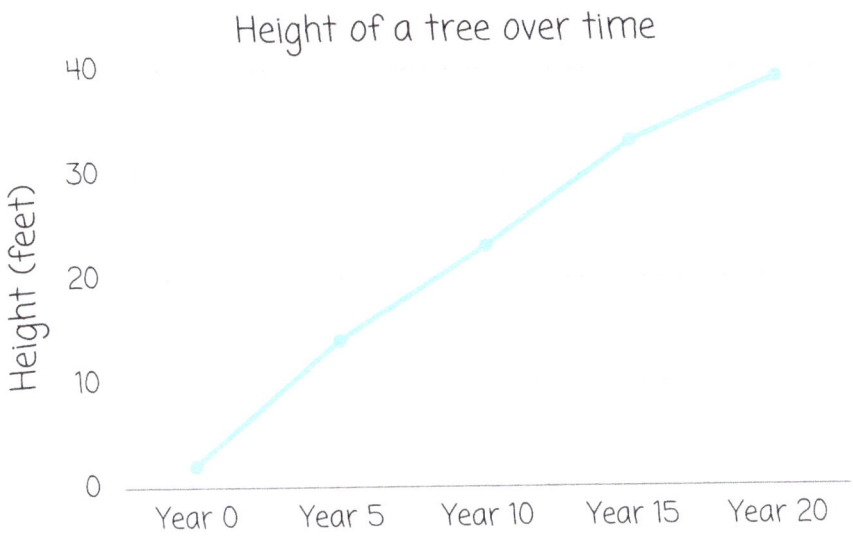

The horizontal part of the chart (the **x-axis**) is labeled with time counts (years, in this case), and the y-axis represents what is being measured at these time points (tree height). Since line graphs are meant to show *changes* over time, it's acceptable to not start the y-axis scale with zero, especially if the measurements are big.

But still be careful here: depending on how the numbers are spaced, a line graph can be manipulated to make it seem like the changes over time are either huge or tiny. For example, if the height of a tree increases by about 3 feet per year, and the numbers on the scale count by 100s, we'd barely be able to see the change from year to year!

Here's another way numbers can be deceptive: have you ever heard on an advertisement that "3 in 4 doctors" or "9 out of 10" doctors recommend a certain medication or healthcare product?

What they're trying to convince you of is that 75% or 90% of all doctors recommend it. But have you thought that maybe these sellers *only asked* four doctors, or *only* ten doctors? And maybe those few doctors were specifically chosen because the seller already knew they would recommend the medication! (Not to mention paid to say they recommend it!)

The point isn't that all these advertisers are trying to trick you, but that you should be careful when you encounter statistics like these.

Ask questions like:

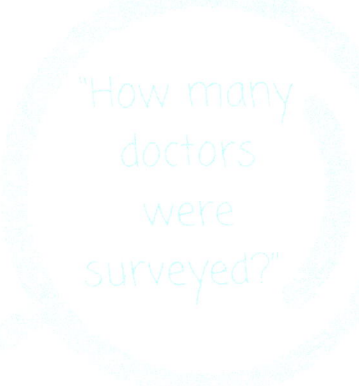

"How many doctors were surveyed?"

"Do the doctors chosen here really represent the views of all doctors?"

If you don't have the answers to these questions, then you can't be sure that a majority of doctors truly recommend the medicine.

Takeaway #8: Be careful when looking at charts or statistics because numbers can trick you!

Where can you learn about scientific research?

Researchers publish papers in scientific journals to report their discoveries. Most of the time, the articles in these journals are read by scientists, not people outside the field of research. But when discoveries are immediately impactful or relevant to the lives of ordinary citizens, then scientists bring their findings to news outlets that the public has access to. So, if you want to be up to date on recent advances in science, check out the science and technology sections of a newspaper!

PART V

SCIENCE IS GREAT BUT NOT PERFECT

Scientists are human, so they make mistakes. As learners of science, we just have to be aware that this means science and scientists might not always be 100% right.

Scientists are experts and know a lot more than most of us about their research, but as they do more investigation and experimentation, they might discover that some things they thought were true at first are not actually.

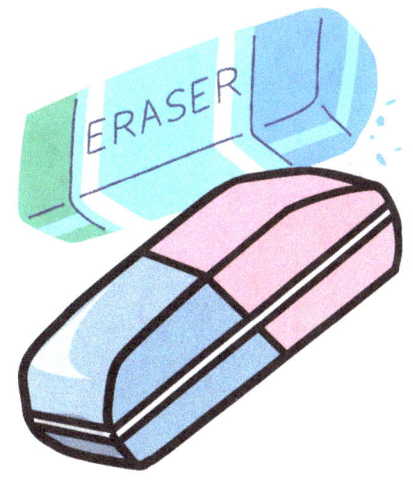

This is why our scientific knowledge is always changing: existing theories are being modified to incorporate new evidence, and new discoveries are being added to it.

Here's an example: centuries ago, many of the brightest minds in the world felt extremely sure that the Sun and the other planets in our solar system revolved around Earth. And because these incredibly intelligent thinkers believed the Earth was the center of the solar system, so did almost everyone else.

But later on, other scientists came along and argued that the Sun was actually the center of the solar system and all the planets (including Earth) revolved around it.

After many years of debate, more observation, more mathematical calculations, and more evidence, people slowly began to accept that the Sun really might be at the center of the solar system and that previous thinkers were wrong.

Today, we have advanced space probes that collect information and images to show us where the planets are in relation to the Sun and how they're moving.

This is a major example of how our understanding of science can change over time.

Sometimes scientists may also make mistakes when they collect data, and this could change the results of their studies. What are some mistakes we might have made when we collected our data about running?

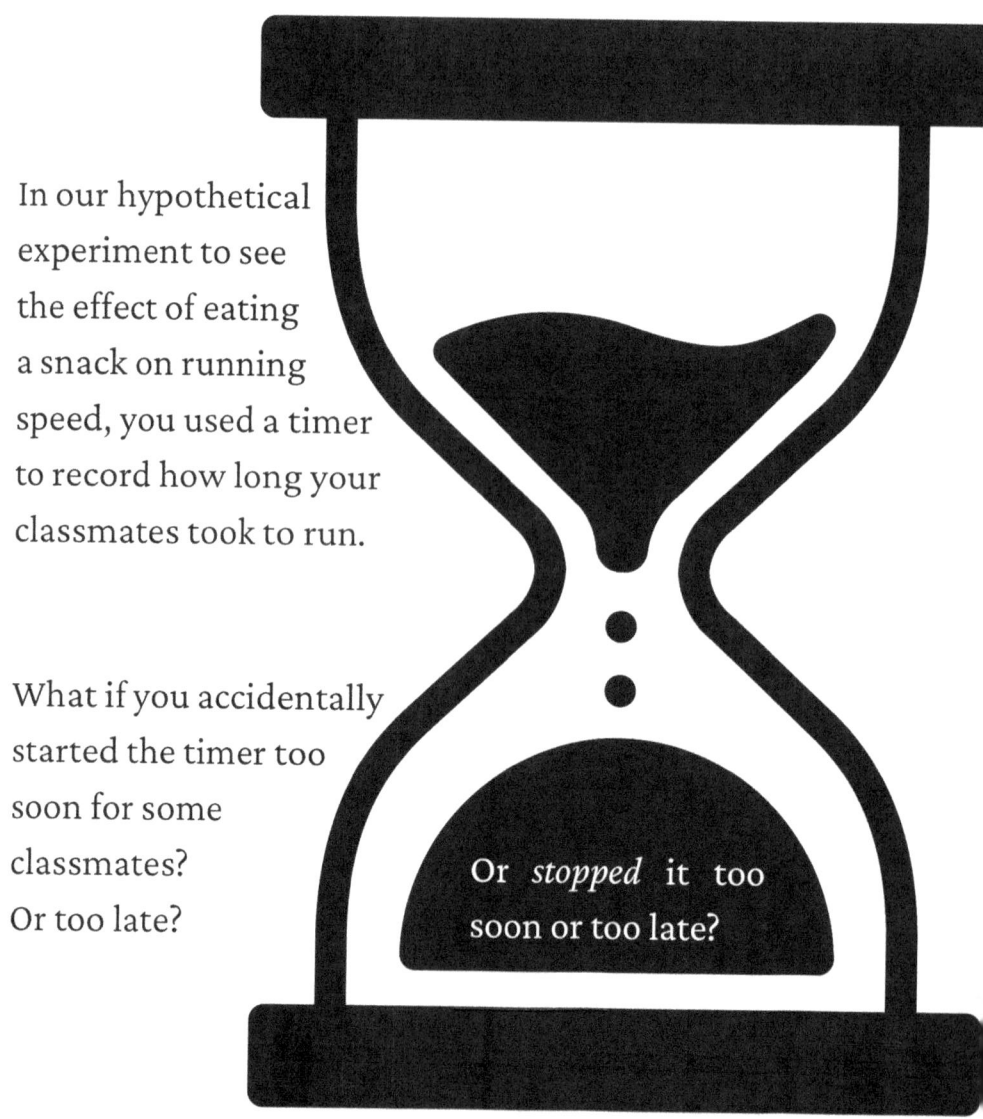

In our hypothetical experiment to see the effect of eating a snack on running speed, you used a timer to record how long your classmates took to run.

What if you accidentally started the timer too soon for some classmates? Or too late?

Or *stopped* it too soon or too late?

In our correlational study where we collected data on the heights of classmates, we might've made mistakes too.

Let's say we measured heights using a meter stick. What if the way we held the stick against some people was tilted or crooked? What if we accidentally read the centimeters side of the stick instead of the inches side for some people?

These are possible data collection mistakes that would make the results of our study less reliable because we would have inaccurate measurements.

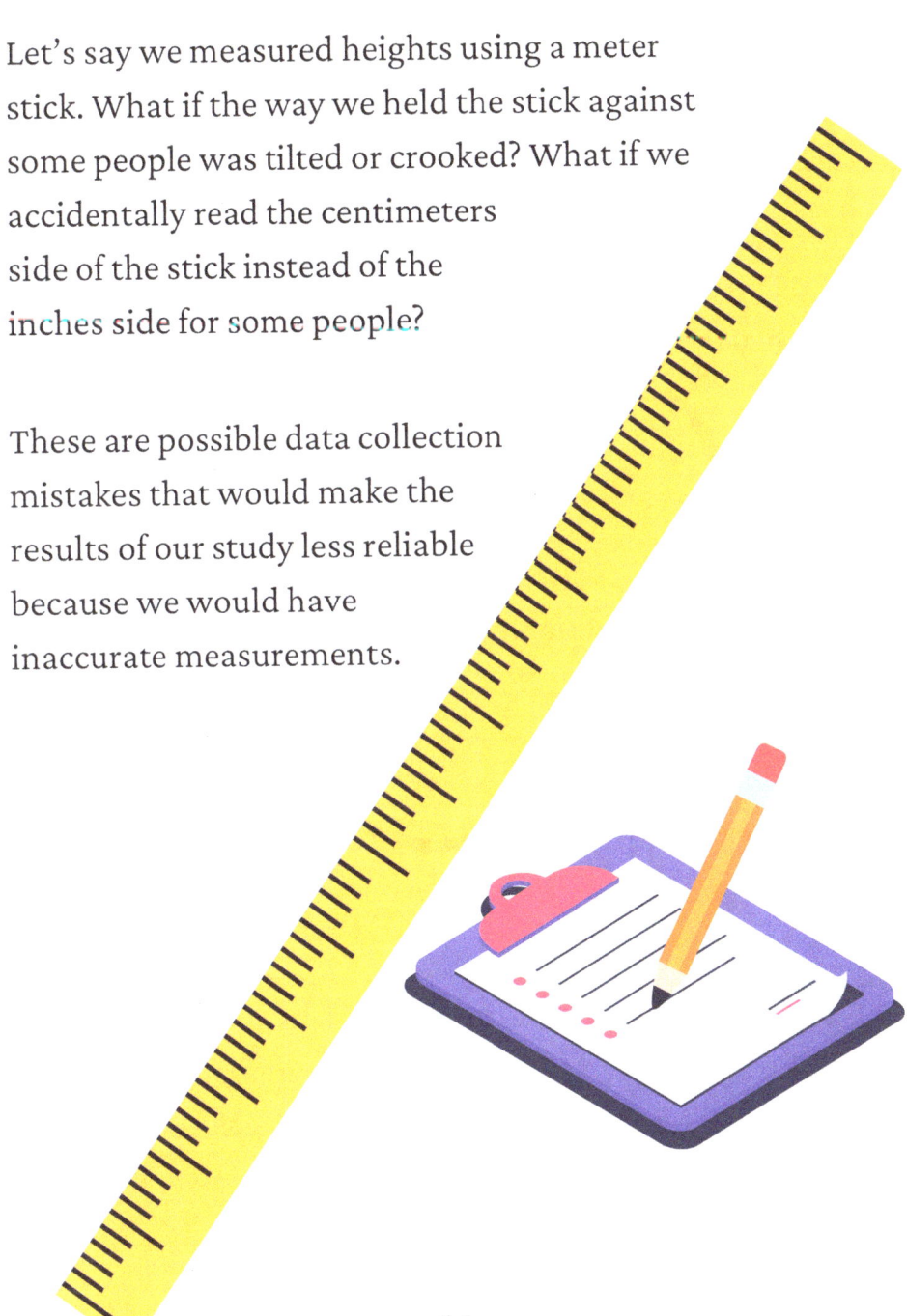

Other times, experiments aren't designed well, leaving room to doubt whether they really back up our hypothesis or claim.

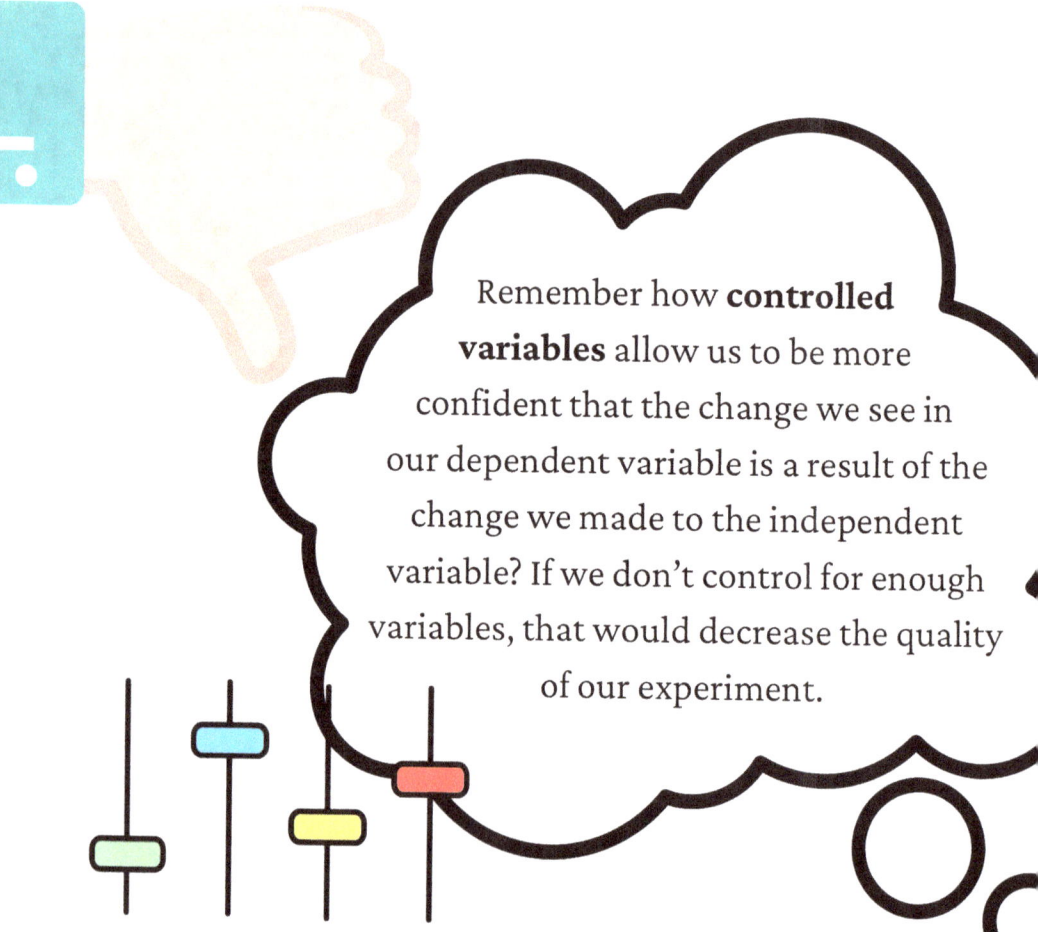

Remember how **controlled variables** allow us to be more confident that the change we see in our dependent variable is a result of the change we made to the independent variable? If we don't control for enough variables, that would decrease the quality of our experiment.

In the experiment on eating a snack and running, if we hadn't tried to make sure that both groups ran at the same time of day, ran the same distance, and wore sneakers or comfortable shoes, that would have been a mistake on our part.

Unfortunately, it isn't possible to control every single variable.

For example, some people might have taken the study more seriously than others and therefore put more effort into running fast.

Or maybe some people had gym class earlier in the day and were more tired than others when they participated in our study.

These factors might have influenced how fast people were able to run, but you can't control how hard people try or how tired they are. The best we can do is keep as many variables constant as possible across our experimental and control conditions.

Another mistake that scientists sometimes make is not testing their hypothesis on enough people, plants, rocks, or whatever it is that they're studying.

Scientific sidenote: Samples

The collection of objects, substances, entities, or people that scientists conduct studies *on* are called samples. Just like you sample food by tasting a bit of it, scientists take soil samples, water samples, rock samples, and samples of people in a population to study.

For example, environmental researchers couldn't possibly study *all* of the soil on Earth. They can only take small containers of it into a lab. The amount of soil these researchers collect to study is called the **sample size**.

If a scientist says that eating chocolate makes people smarter, and he or she only tested it on three people(a tiny sample size!), this might not be a reliable finding.

Maybe the three people really did score higher on a test after eating lots of chocolate, but if it's only three people being tested, it could've been a complete coincidence.

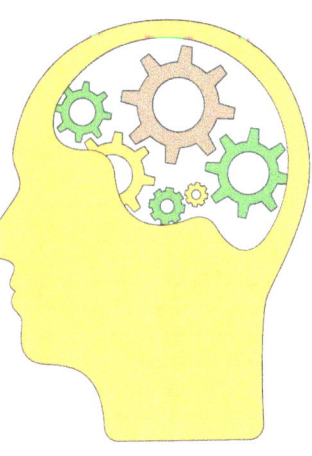

Similarly, if a scientist says that Fertilizer X makes plants grow faster and he or she only tested it out on three plants, that's not strong enough evidence.

Using a large sample size makes our results more reliable.

But a larger sample size means more resources, which is more expensive.

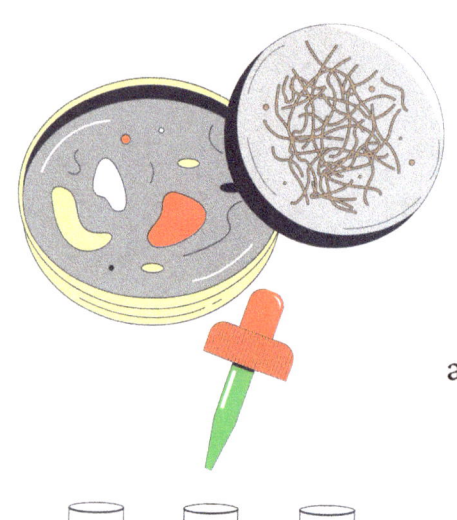

Depending on what a scientist is studying, it's not always possible to use a large sample size.

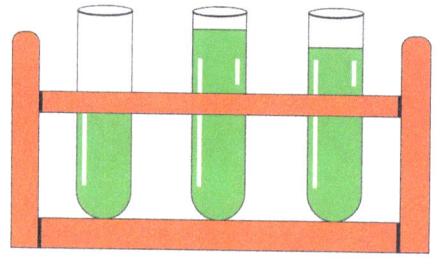  

What if some scientists want to study diamonds? Diamonds are very expensive, so it doesn't make sense to buy a ton of them to test on.

And what if scientists want to study some huge process like how Earth's oceans rise and change temperature over many years?

They can't possibly experiment by changing the temperatures and conditions of entire oceans! Instead, they can build models—for example, mini versions of oceans using big containers of water and ice—that they test on by, for example, heating up to mimic the effect of the sun. The bigger and more accurate scientists build such models, the more it costs.

Scientists don't have unlimited resources and money to use on research, so they can't be perfect with their studies.

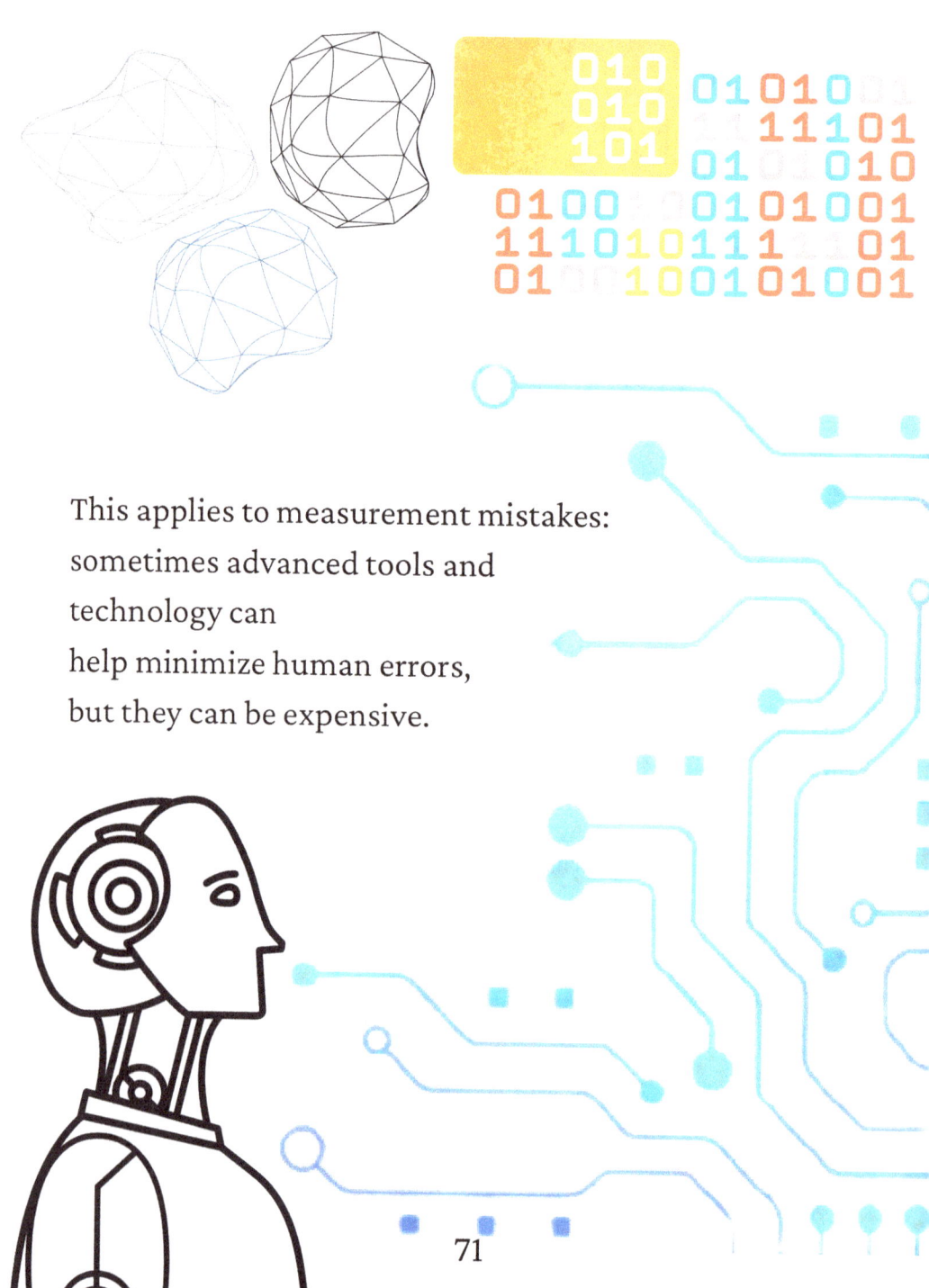

This applies to measurement mistakes: sometimes advanced tools and technology can
help minimize human errors, but they can be expensive.

> **Takeaway #9: Scientific research isn't always completely accurate because scientists make mistakes and don't have the time, money, and resources to conduct perfect studies.**

Scientists making mistakes or being unable to design perfect studies is one reason why replicating studies in science is so important!

If the first time we run a study we get one result, but the second, third, fourth, and fifth time we get a completely different result, we might suspect that our first result was due to a human error. But the more times we repeat a study and do get the same outcome, the more confident we can be that it's because the finding is real and not due to a mistake.

Even with its limitations and room for mistakes though, science is still an incredibly useful, informative, and reliable method that we have to learn about the world. Even if we can't entirely prove anything for sure, science allows us to get as close as we can.

Science has been critically important to us, and will continue to improve human society through...

medicine, **engineering**, **technology**, **astronomy** (finding other planets that can sustain life), **agriculture** (growing food to feed the increasing population), **psychology** (understanding our brains),

...and much more than we can imagine!

Because science is everywhere, it's important that you understand how it works. Now that you've read this book, you're off to a great start!

> **Takeaway #10: Understanding what science is and how it works is more valuable than memorizing scientific facts.**

Glossary/Index

Research (p. 2) - the general process by which scientists make discoveries about the natural and physical world

Scientific method (p. 7) - the research process of asking questions, making hypotheses, and testing them using observation and/or experimentation

Observation (p. 8) - the act of noticing, sensing, or perceiving something to exist

Research question (p. 9) - a step of the scientific method; a question about the world that scientists attempt to answer through observation or experimentation

Hypothesis (plural: hypotheses; p. 10) - a testable prediction about the answer to a research question. It predicts the outcome of a study or experiment

Experiment (p. 11) - a form of scientific investigation that involves changing one factor/variable to see if it causes a change or difference in another factor/variable

Procedures (p. 11) - the precise steps that are involved in an experiment. It's important to have a step-by-step procedure in experiments so that future scientists can follow it to re-conduct the experiment.

Evidence (p. 12, 45-46) - information, data, or observations that can be used to justify (but not "prove") a scientific claim

Independent variable (p. 13) - the factor in an experiment that the researchers/scientists themselves manipulate and expect to cause a change in something else

Dependent variable (p. 13) - the factor in an experiment that we expect to be changed or influenced *by* the independent variable; the outcome we measure

Control condition (p. 14) - the group or trial in an experiment that we leave in its natural baseline state; not manipulated or given a treatment

Replication (p. 15) - repeating studies and experiments to see whether the same results are consistently obtained each time

Variable (p. 16) - any factor in a study that can change

Controlled variables (p. 17) - the variables or factors that we keep constant across control and experimental conditions

Correlational/observational study (p. 23-24) - studies where we collect data for two or more variables and determine whether there is a relationship between the two. There may be a **positive correlation** (p. 25), where values of one variable increase as the values of another variable also increase, or there may be a **negative correlation** (p. 25), where values of one variable decrease as the values of another increase.

Confounding variables (p. 26) - variables we don't collect data on that make it difficult to know whether there is a cause-effect relationship between two other variables. We might think X causes Y, but it might be that confounding variable Z causes *both* X and Y.

Scatter plot (p. 30) - a visual display of data that plots pairs of values on a coordinate plane. If all the dots form roughly a straight line, then there is a correlation between them.

Theory (p. 36) - a scientific explanation of events, processes, and things in the world that logically brings together many facts and pieces of evidence.

Phenomenon (plural: phenomena; p. 37) - things, processes, or events in the physical world that scientists study

Theory of evolution (p. 37) - an explanation by Charles Darwin that explains how species of organisms change genetically over time to adapt to their environment

Fact (p. 41) - observations and findings in science that are supported (and not contradicted) by all available evidence and usually accepted as true by nearly all

Proof (p. 42) - establishing something as absolutely certain; a word that should not be used when talking about science

Confirmation bias (p. 46) - the harmful tendency to search only for information that backs up our beliefs and hypotheses while ignoring contradictory information

Peer review (p. 49) - reading one another's research papers before publication to ensure that the research was carried out properly and findings are conveyed clearly and accurately

Bar charts (p. 53) - a type of data display that represents values as bars (the higher the bar, the larger the value); useful for comparing values across different groups or categories

Line graphs (p. 56) - a type of data display that represents values as points plotted at different heights (the higher the point, the larger the value) can connected to show trends and changes in value over time

Sample (p. 57) - the pool of objects, substances, or subjects that a researcher studies. The larger the sample size, the more reliable the findings are.